LABORATORY NOTEBOOK

www.ingramcontent.com/pod-product-compliance
Ingram Content Group UK Ltd.
Pitfield, Milton Keynes, MK11 3LW, UK
UKHW050634081025
8288UKWH00043B/1483

9 781681 451558